SCIENTIFIC AMERICAN EDUCATIONAL PUBLISHING

SCIENTIFIC AMERICAN INVESTIGATES LIFE CYCLES

THE STRANGE LIFE CYCLE OF A SALMON

ANNA MCDOUGAL

Published in 2025 by The Rosen Publishing Group
in association with Scientific American Educational Publishing
2544 Clinton Street, Buffalo NY 14224

Portions of this work were originally authored by Mark Harasymiw and published as *The Bizarre Life Cycle of a Salmon*. All new material in this edition authored by Anna McDougal.

Library of Congress Cataloging-in-Publication Data
Names: McDougal, Anna, author.
Title: The strange life cycle of a salmon / Anna McDougal.
Description: Buffalo, NY : Scientific American Educational Publishing, [2025] | Series: Scientific American investigates life cycles | Includes bibliographical references and index.
Identifiers: LCCN 2024011785 (print) | LCCN 2024011786 (ebook) | ISBN 9781725350977 (library binding) | ISBN 9781725350960 (paperback) | ISBN 9781725350984 (ebook)
Subjects: LCSH: Salmon–Juvenile literature. | Salmon–Life cycles–Juvenile literature. | Life cycles (Biology)–Juvenile literature.
Classification: LCC QL638.S2 M33 2025 (print) | LCC QL638.S2 (ebook) | DDC 597.5/6156–dc23/eng/20240320
LC record available at https://lccn.loc.gov/2024011785
LC ebook record available at https://lccn.loc.gov/2024011786

Designer: Tanya Dellaccio Keeney
Editor: Caitie McAneney

Photo credits: Cover Pukhov K/Shutterstock.com; p. 5 (top) Krasowit/Shutterstock.com; p. 5 (bottom) Ronnie Chua/Shutterstock.com; p. 7 slowmotiongli/Shutterstock.com; p. 9 J. Helgason/shutterstock.com; p. 11 (top) Kevin Cass/Shutterstock.com; p. 11 (bottom) Mark A. McCaffrey/Shutterstock.com; p. 13 Canadapanda/Shutterstock.com; p. 15 Anatoliy Lukich/Shutterstock.com; p. 17 Crystal Kirk/shutterstock.com; p. 19 CrackerClips Stock Media/Shutterstock.com; p. 21 Andrei Armiagov/Shutterstock.com.

Printed in the United States of America

CPSIA compliance information: Batch #CSSA25. For Further Information contact Rosen Publishing at 1-800-237-9932.

CONTENTS

Words in the glossary appear in **bold** type the first time they are used in the text.

A LONG WAY HOME

A salmon's life is truly a wild journey. They are anadromous (uh-NA-druh-muhs). This means they're born in fresh water, spend their adult lives in salt water, and then return to fresh water to **spawn**. This journey is not only long, but also full of challenges.

After hatching in freshwater, salmon journey to the ocean. They stay for a bit, but it's not where they end their lives. Instead, they go on another journey. This time, they head home. It's an upstream swim for the strong salmon. Let's follow the salmon's strange life cycle!

FUN FACT

WHEN A SALMON FINDS AN OBSTACLE IN THEIR PATH, THEY OFTEN JUMP OVER IT!

Salmon start their lives as eggs in a body of fresh water, such as a lake or stream.

SALMON SPECIES

Salmon are found in both the Atlantic and Pacific Oceans. Seven species of Pacific salmon exist, some in North American waters and others in Asian waters. Species include the Chinook, chum, coho, pink, and sockeye salmon.

The Atlantic salmon is found throughout the North Atlantic. In European waters, it's found as far south as Portugal. In North American waters, it's found as far south as Maine. In 1956, pink salmon were brought to the Great Lakes, possibly by accident.

There's only one Atlantic salmon species.

SMALL FRY

Female salmon lay their eggs where they were born years earlier. Each female can lay thousands of eggs in the **gravel** of the stream or lake. The young salmon, called fry, hatch after several weeks or months. The fry stay in the gravel for a time, living off their **yolk sacs**.

Some species swim to the ocean quickly. Others remain for a while to feed on **plankton**, insects, and **crustaceans**. Young Atlantic salmon usually remain where they were born for one to two years.

In some places, young Atlantic salmon remain in their home waters for up to 8 years before swimming to the ocean!

FUN FACT

GROWING SALMON ARE CALLED PARR.

ADULT SALMON

Salmon swim downstream to the ocean. Then their bodies change to help them live in salt water. These changes are called smolting.

While in the ocean, adult salmon feed on small fish, water bugs, and crustaceans. They may stay in the ocean for as few as two years, like pink salmon. They may stay for as many as 7 years, like Chinook salmon. Salmon spend their time in the ocean eating to prepare themselves for their long journey to spawn.

CHINOOK SALMON

Chinook, or king, salmon grow up to 58 inches (147.3 cm) and 126 pounds (57.1 kg). They're the largest salmon species.

PINK SALMON

FUN FACT

PINK SALMON ARE THE SMALLEST KIND OF SALMON, WEIGHING ONLY A FEW POUNDS.

RETURNING HOME

The salmon's journey back to its home waters is deadly. Few survive it. Only about one in every 1,000 salmon eggs become adults and return to their home waters. Some are eaten by other animals. Some die of sickness. Some die on the return journey because they haven't stored enough fat.

Salmon use their sense of smell to recognize their home stream. Once at the spawning grounds, male salmon fight other males to **mate** with the females. Female salmon fight with other females for a good place to lay eggs.

The journey back home is very challenging because it is upstream!

FUN FACT

SOME SALMON CHOOSE DIFFERENT LOCATIONS TO SPAWN FROM THEIR HOME WATERS. THIS IS CALLED STRAYING.

THE NEXT LIFE CYCLE

Female salmon make four or five nests when they're ready to lay eggs. These nests are called redds. The female uses her fins and tail to make a shallow hole in the gravel. After she lays her eggs in the redd, a male salmon adds cells called milt to them.

The female and male do this several more times. In most species of salmon, the male and female die after all the eggs have been spawned. Their life cycle is now complete, and a new one will begin.

The upstream journey of many salmon each year is called a salmon run.

FUN FACT

FEMALE SALMON COVER THE EGGS IN THEIR REDD WITH GRAVEL TO KEEP THEM SAFE.

SALMON AT RISK

The journey to and from the ocean is risky. That's because plenty of animals would love to eat salmon! Many salmon are caught by bears, otters, and eagles. Eating salmon is a great way for bears to fatten up before they sleep through winter.

Salmon have become **endangered** in some areas and completely wiped out in others. Increases in water **temperature** and water pollution can also harm salmon. This will get worse as water temperatures continue to rise due to **climate change**.

Many people fish for salmon, such as this fly fisherman.

FUN FACT

HYDROELECTRIC DAMS ACROSS RIVERS CAN MAKE SALMON'S JOURNEY HARD.

HOW CAN WE HELP?

There are things that people can do to help salmon on their journey. Some people have helped salmon by setting up hatcheries. Hatcheries are safe places for young salmon to live and find food, water, and space.

To help salmon swim past dams, people have built fish ladders. A fish ladder is a series of pools that salmon jump up into so they can swim up and around the dam. They're somewhat like steps, but for salmon! They help solve a human-made problem.

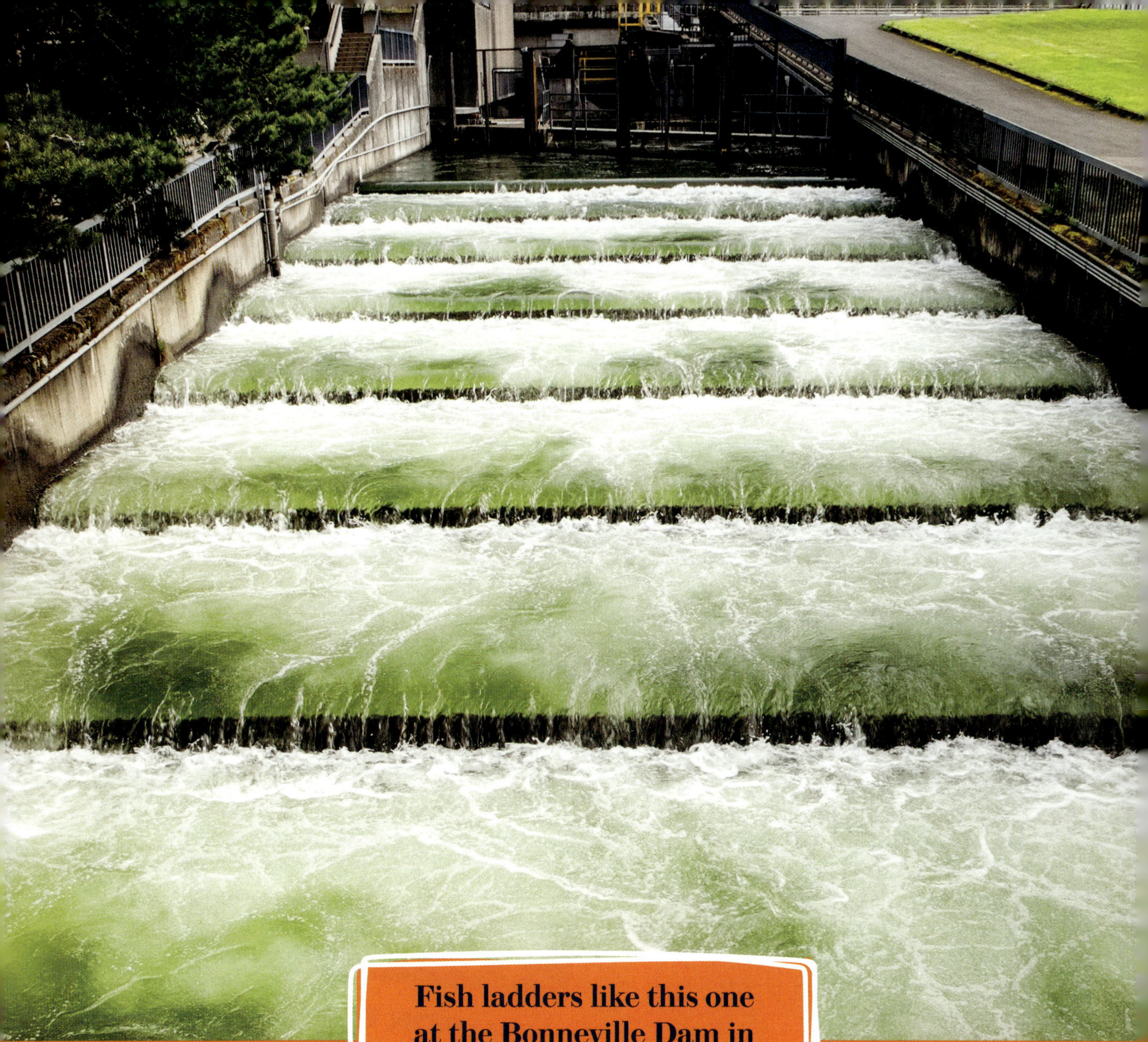

Fish ladders like this one at the Bonneville Dam in Oregon help salmon migrate, or move from place to place.

FISH FARMS

Salmon are the second-most popular seafood eaten in the United States. Their meat is healthy and easy to cook. Unfortunately, they are overfished in the wild. Because of this, many fish are raised on salmon farms.

The more fish are eaten from salmon farms, the fewer wild salmon have to be caught. In fact, more salmon live on farms than in the wild. People raising salmon must first understand their strange life cycle. These fantastic fish are one of a kind!

FUN FACT

SALMON FARMS HAVE DOWNSIDES. THEY CAN LEAD TO ILLNESS IN THE FISH, POLLUTION, AND SALMON ESCAPING INTO NEW WATERS.

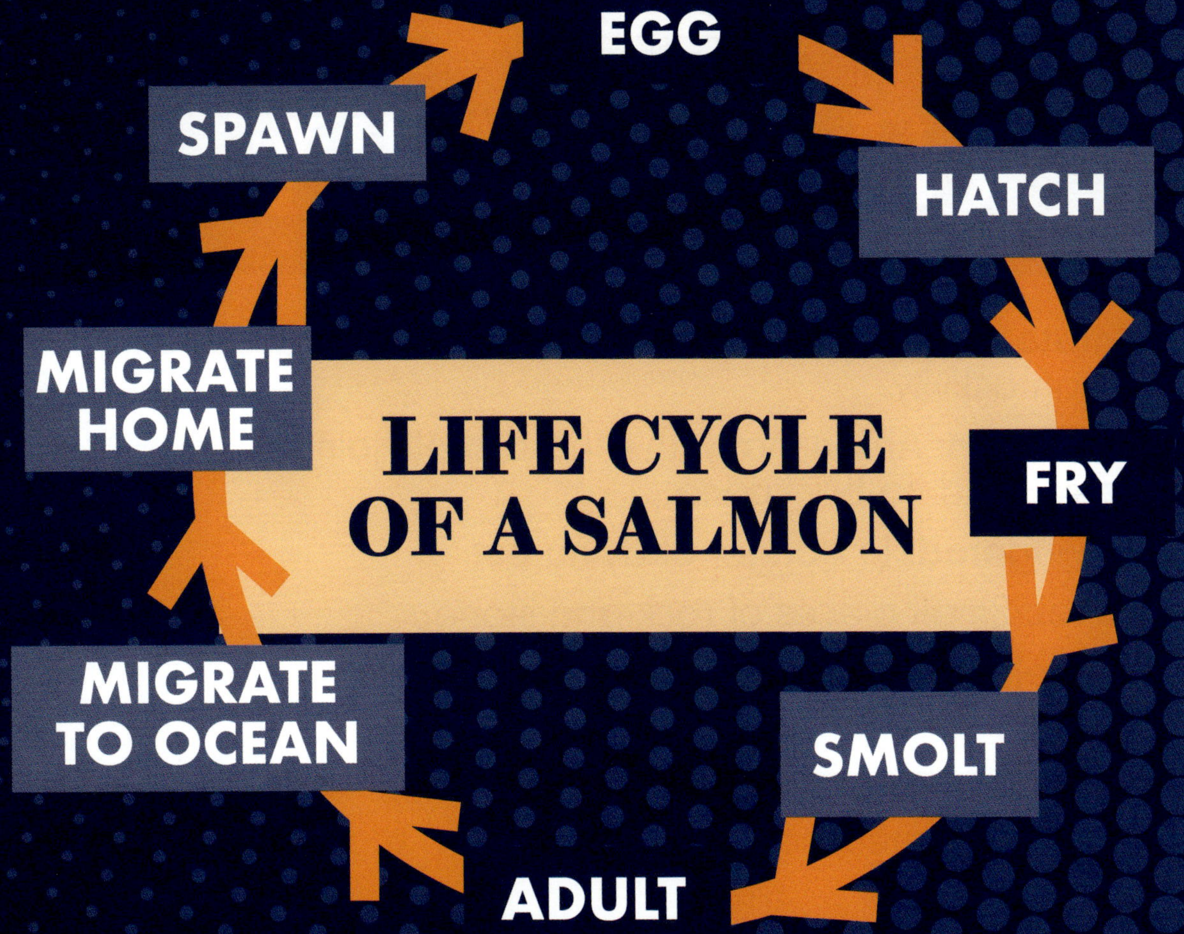

Young farm salmon are raised in fresh water and then moved to salt water cages, like these, when they become adults.

GLOSSARY

climate change: Change in Earth's weather caused by human activity.

crustacean: An animal with a hard shell, jointed limbs, feelers, and no backbone.

endangered: In danger of dying out.

gravel: Small pieces of rock and stone.

hydroelectric: Describing something that produces electricity from the motion of water.

mate: To come together to make babies.

obstacle: Something that stops forward movement or progress.

plankton: Tiny plants and animals floating in the water.

spawn: To produce young, usually in large numbers.

temperature: How hot or cold something is.

yolk sac: A supply of food surrounding the young of some kinds of animals.

FOR MORE INFORMATION

Books

Holmes, Kirsty. *Life Cycle of a Salmon*. New York, NY: KidHaven Publishing, 2022.

Rice, Jamie. *A Salmon's Life Cycle*. Minneapolis, MN: Jump!, Inc., 2023.

Websites

Facts on Fish

kids.nationalgeographic.com/nature/article/facts-on-fish
Learn more about fish and the challenges they face.

Salmon

kids.britannica.com/kids/article/salmon/353740
Explore more facts about salmon, such as where they live and their features.

Publisher's note to educators and parents: Our editors have carefully reviewed these websites to ensure that they are suitable for students. Many websites change frequently, however, and we cannot guarantee that a site's future contents will continue to meet our high standards of quality and educational value. Be advised that students should be closely supervised whenever they access the internet.

INDEX